DATE DUE

634.9　　$13.00
PEL　　Pellowski, Michael
　　　　What's it like to be
　　　　a Forest Ranger

DISCARD

HACIENDA MAGNET SCHOOL
1290 KIMBERLY DRIVE
SAN JOSE, CA 95118

What's it like to be a...
FOREST RANGER

Written by Michael J. Pellowski
Illustrated by George Ulrich

Troll Associates

Special Consultant: Robert Conklin, *U.S. Forest Ranger, New York State.*

Library of Congress Cataloging-in-Publication Data
Pellowski, Michael.
    Forest ranger / by Michael J. Pellowski; illustrated by George Ulrich.
    p.   cm.—(What's it like to be a...)
    Summary: Outlines the various jobs performed by forest rangers, including firefighting, inspecting trees for disease and insect damage, and keeping track of animals and the water supply.
    ISBN 0-8167-1422-3 (lib. bdg.)   ISBN 0-8167-1423-1 (pbk.)
    1. Forest rangers—Vocational guidance—Juvenile literature.
[1. Forest rangers. 2. Occupations.] I. Ulrich, George, ill.
II. Title. III. Series.
SD387.F6P45 1989
634.9—dc19   88-10355

Copyright © 1989 by Troll Associates, Mahwah, New Jersey
All rights reserved. No part of this book may be used or reproduced in any manner whatsoever without written permission from the publisher.
Printed in the United States of America.

10 9 8 7 6 5 4 3 2 1

What's it like to be a...
FOREST RANGER

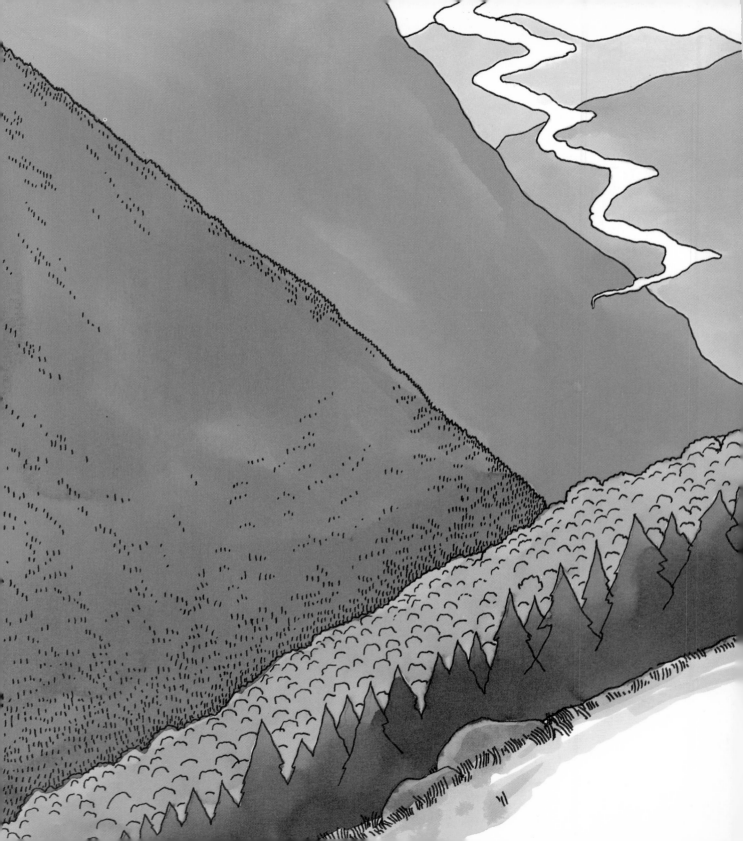

A thin stream of smoke is rising from above the trees. Woods are burning!

High on a mountain top, someone sees the fire. She is Ellen Hanley, a lookout for the National Forest Service. Using a special instrument, Ellen pinpoints the fire's location. Then she alerts the forest rangers.

Direction Scale

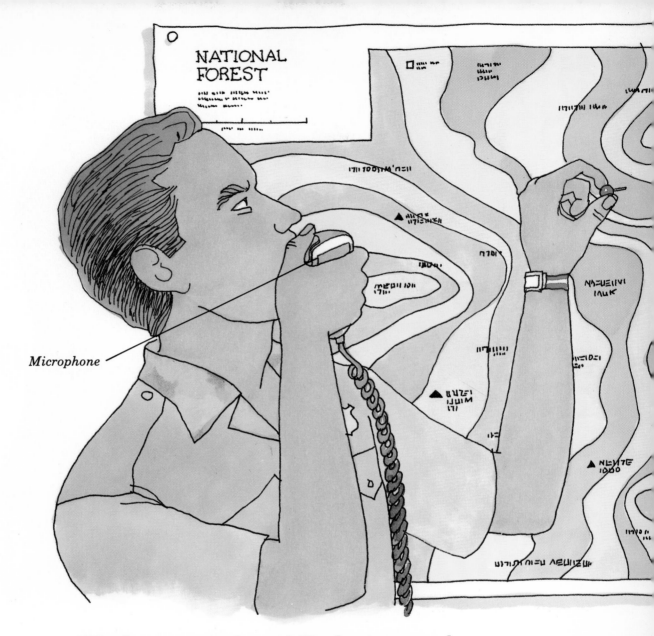

Microphone

"Hanley to ground crew! Hanley to ground crew!" she calls into her two-way radio. "Fire burning! Fire burning!"

Jim Dayton, the head ranger, answers the call. He marks the fire's location on a map. Then he sends the fire-fighting crew into action.

The fire—started by some dry leaves, twigs, and grass—is spreading fast. The ranger yells to the crew chief, "Let's make a firebreak, before this fire starts climbing up the trees!"

A firebreak is a special way to fight a forest fire. The rangers form a line. They go to work, removing anything that might burn. They use shovels, chain saws, and other hand tools. When the fire reaches the firebreak, it cannot spread any further.

For big fires, rangers use bulldozers to make firebreaks. They also use airplanes and helicopters to reach fires that are burning out of control. Special fire fighters parachute down to fires that ground crews cannot reach. They are called "smoke jumpers."

11

After the fire is out, the crew returns to the ranger station.

"Good work, Ellen," says Jim Dayton to the lookout.

"Thanks," Ellen replies.

Ellen introduces the rangers to a camper named Bob.

"Bob's been watching," Ellen says. "He wants to be a ranger someday."

"I want to fight forest fires, just like you," Bob says to the rangers.

The head ranger smiles.

"Forest rangers do a lot more than fight fires," he says. "Each ranger has special training to do many jobs."

"Will you tell me all about forest rangers?" Bob asks.

"Sure," says Jim. "Let's go into the ranger station."

"All rangers must go to college to learn about forestry," Jim begins. "Then they are trained in one of many fields."

The head ranger pauses.
"Do you like the outdoors?" he asks Bob.
"You bet!"
"Well, that's a good start," says Jim.
"Rangers love the outdoors!"

Some rangers specialize in tree diseases and insect control. They inspect forests for signs of trouble. It's their job to make sure the trees stay healthy.

Back Pack

Sometimes, a ranger may spot bark beetles in the trees. These insects eat through the tree bark and cause disease. Trees with bark beetles have to be sprayed.

Bark Beetle

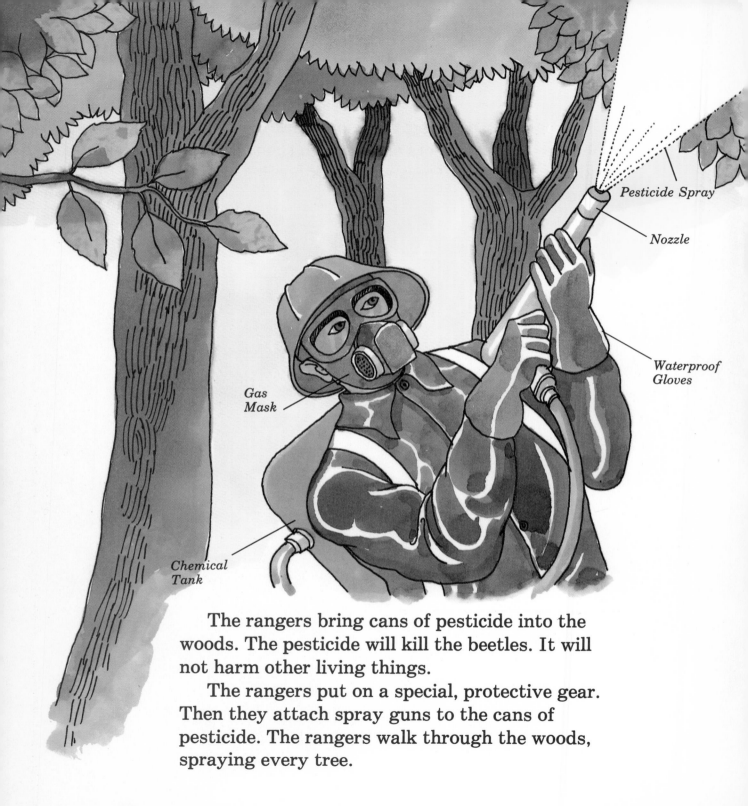

The rangers bring cans of pesticide into the woods. The pesticide will kill the beetles. It will not harm other living things.

The rangers put on a special, protective gear. Then they attach spray guns to the cans of pesticide. The rangers walk through the woods, spraying every tree.

Healthy trees are often used for lumber. These trees are called timber. Timber control is the job of other rangers. Timber control means that some trees must be cut down to allow other trees to grow.

Timber

Spare Tire

Four-Wheel Drive Jeep

Spade

Extra Fuel Tank

Map

Rangers decide which trees to cut. They measure the trees to see how much lumber there will be. Then the trees to be cut are marked with white paint.

Spray Paint

Tape Measure

"Do rangers cut the trees themselves?" asks Bob.

"No," says the ranger. "Trees used for timber are sold to timber companies. Company lumberjacks do the cutting."

Jim takes Bob on a ride through the forest. "Many rangers specialize in wildlife," says Jim. "They make sure wild animals have a good home in the forest. They plant trees that give the animals food and shelter."

Seat Belt

Four-Wheel Drive Jeep

"Look," says Bob. "There's a mother bear and her cubs."

"Each one is tagged and has a number," says the head ranger. "Keeping track of forest animals is part of the job, too!"

Some rangers are in charge of grazing land for animals. This is land on which animals eat grass and roam free. In the western United States, the government owns lots of land. Ranchers pay the government to let their animals graze on it.

Cattle and sheep graze on the land. Wild animals eat the grass, too. Rangers make sure there is enough food for all the animals. They also check the soil to make sure it is in good condition.

"The most valuable resource in our forests is water," says Jim, pointing to a waterfall. "Rangers must pay attention to the water needs in their area. They use special instruments to measure snowfall and how much snow melts."

The melting snow tells the rangers how much water to expect on the ground. Water is important for growing things.

Weather Instruments Center (Humidity, Air Pressure, Temperature)

Snow Moisture Gauge

Snow Depth Gauges

"Sometimes, snow builds up on mountain sides," Jim tells Bob. "This may cause an avalanche, or serious slide. Rangers use explosives to make the snow slide harmlessly down the mountains."

BOOM! WHOOSH!
Jim and Bob watch some snow slide gently from the mountain top.
"The mountain is safe now," says Jim.
Bob sees two rangers ski away.

"Wow!" says Bob. "I had no idea that forest rangers do all those things!"

"Do you still want to be a ranger?" Jim asks with a smile.

"More than ever," answers Bob.

"Well," says Jim, putting his ranger's hat on top of Bob's head, "maybe someday you will."